THE UNKNOWINGNESS

THE UNKNOWINGNESS

PAST, PRESENT, AND FUTURE DIVULGES REELING YOUR HEADS

AMELIA HATHOW

ISBN: 978-1-958150-92-4
The Unknowingness: Past, Present, and Future Divulges - Reeling Your Heads

Paperpack

September 2022

Subjects:
SCIENCE / Chemistry / Electrochemistry
PSYCHOLOGY / Emotions
BODY, MIND & SPIRIT / Parapsychology / Out-of-body Experience

ameliahathow.com

dedicated to humanity
to make an effort
to make a change
for the good
let us come together
now

or else

TABLE OF CONTENTS

Introduction

WHAT CAME BEFORE the big bang? God. Not only God, but a streuth full of teeming life that is still iridescent in color, not to be seen by any naked eye. Seriously though, the Big Bang was one of many throughout our lifetime. Me, being the speaker for God itself – Tola (a name that changes over time like everything else. I am someone else now as this writer writes more words by MY doing, MY will over him/her). You see, I am an alien who this writer can feel in such a way that hearing becomes a misnomer who is only feeling at his first layer of skin, what I say and simply typing it out, verbatim – The Truth. Your truth. Our truth. We didn't come from One Big Bang. No. We, your earlier and much older version of "somewhat" yourselves actually came from 15 other such similar Big Bangs throughout lifetimes. 15, at this point in time and space, to be somewhat precise with our calculations and that is how fast your lives are compared to ours. Meaning, for every one of yours we are 15 less years than you. So, our lives are much much longer than yours. 15 also, because that is the nth degree of our differences and is very scientific indeed. Take for instance, your leap year.

We will have 15 of them in one decade instead of yours every so often, not to be so precise, but actually yours is every 4.569 leaped year. So, you all have been living a "lie" only because of the inaccuracies of measure not to be taken lightly for this "lept" year does have a significant effect on growth rates of animals, plants, molecularity of water based molecules and so much more. It all has to be precisely measured in order for real progress to happen. Your Big Bang actually occurred 34 KA-billion years ago. We have KA in front because that is what burst "out" first, potassium and an alkaline. The alkaline was soluble in potassium, seeing that your Big Bang was in the form of a liquid, just like when babies are born "in" a liquid. The same went for you all, in a liquid form first, then gas, then a much less dense version of gas, then the real explosion happened forming solids. So, yes, there were two "explosions" for those of YOU who kind of suggested that to those very few who even considered such a thing. So, there you have it, your Big Bang theory and some tidbits of knowledge to look at especially the K-A part. Look, another clue to what the alkaline is, negative polarity of an a (not capitalized, to your liking).

Prelude to destruction will be someone to come. Not Christ but the afterbirth of him. What is that exactly, right? Right you are to think you all are not THE AFTERBIRTH OF SUCH A THING AS CHRIST because that is ALL he is, your birth waste spreading disease.

From us to you, with love, mankind and women alike.

OUR Christ Agenda

THE POINT OF eternal"ness" is not praying to someone who can't take away what was never his in the first place. Christ was merely a figment of our imaginations. A figurehead that every one of your kind could revere. An example of sacrifice. Sacrifice being the key term. He did sacrifice himself and he went above and beyond the call of duty. In fact, what he was actually doing was paving the way on what to do with your lives and what needs to be accomplished at a personal level in order to achieve what he deserved, to be hung from two "crossed" pieces of wooden planks. It's not the cross that killed him. It was the sorrow people felt for him. The woes and the tears pleading for his life, to be spared for he was their king. A king only, though. Never a God. Why would anyone revere a king when they all are mostly God by design? That is reverse thinking. God makes up a sum total of 84% of your total construct, which does include the majority of your spiritual, physical, and mental bodies with a small deviation here and there. But who is really keeping track, for those of you who are so concerned with percentages and who gets what? Not even the greatest of souls would ever consider themselves primarily God.

Once the Lord becomes the revered, the trust of Kings, the KING ITSELF becomes null with ever darkening stains slowly being blotched out of our "mental" existence. IT IS such a place where OUR GOD reigns. Deep in the afterthoughts of those who keep feeding off others' opinions about what Jesus was, your savior, which has never been the case for as long as time began. Instead, something else happened that is true to OUR nature. Life already existed long before even the spark of the fly that began even before mankind began. A type of FLY existed even before Christ was ever a thought to anyone. A simple little fly that likes to feed off the dust and grime of those who never like to clean their flesh. Yet, people want to believe in someone who gave birth to a fanatical uprising of sorts, killing thousands of non-believers. The crusaders WERE the flies that fed off the dirt and grime of those who ran for their lives in full knowledge that Christ WAS a figment of someone's imagination only to become a significant mental image burned into people's minds. So scared that listening to others isn't an option anymore. Flies buzzing around carcasses of their own savage beatings. Brutalities arised to mass graves only to find flies upon other human flies buzzing and swatting at each other while they look upon their wasted efforts. Only, to soon find out, that Christ will never come back. Because he can't. I, your messenger, an "Alien," CAN. No other will rule your destinies. You have no idea of the real powers we have. Only time and patience precludes our actions to make mockery of you all. Forgive yourselves soon. Time is a-ticking. Tick Tock. Tick Tick Tock. Bellies woe as they read these sinful, piously, outlandishly spittle slathering waste down its own belly fat inside as we prepare your plight of all plights that has never been. Plights after plights, this will be the one to watch if you will even be able to keep your eyes shut long enough before the flames burn them out of their calloused riddled sockets, ablaze goes the first trumpet. Sorrows will follow upon sight of our opening acts. Trust us with these words staring back at you knowing and undoing all of your sinful ways, rests upon your slithering, snaked huskers. Abound. Lay thee upon thy withering ways of last year's tolerances. Only those with desires to love will ever withstand this tremor coming through YOUR clouds, not OURS. You made us do this. YOU asked for it. We only get called when those have gone far enough. See your clouds take different shapes. That is A sign of coming times spinning round and round and round until All is torn up from its roots. Land, homes, people, animals, and, best of all, those sinful preachers who always heeded their own tongues lashing out lie after lie after lie. You all know each other and then, yourselves. Every last one of you will be plucked out of your beds, fields, and skins, rapping against the earth as you fall hundreds and even thousands of feet to your deaths. Some we will let live. Only those who have redeemed themselves enough to be willing to move forward and never backwards again. Watch the trench warfare as those spared squander for food and water while those laying bloodied on the ground will not be saved. Instead, those who are saved will be thrust into slaying the ones laying in their own pile of broken bones. Those will be extinguished with willful feelings of duty pulling the triggers.

Dark days. Dark nights. Dark periods of light in the night, crawling, creepily around the fabled tales as you ALL have been spoonfed. Spoonfed in such ways that even the truth tellers get off the beaten path. Where to go now, they ask? Only THE truth can release us ALL from THE light, burning our skin in great fashion. Fashionable burnt skin. THAT will be your new attire. Burnt skin, ALL around. THIS FABLE, HOWEVER, IS NOT A LIE. The burning of flesh doesn't happen in a false hell. It will happen in your daytime, nighttime, and ALL time UNLESS WE all stop it from happening two years from back when? A month ago. Tell time the way you want to. Time doesn't mind. Neither do we.

Come as they will come to those who seek not only help but healing. Physical impairments will be astronomical in turn. Meaning, both in severity. Being a severity of such proportions that you will need a lot of simple healing supplies to accommodate the wandering wounded. We are talking about hard felt injuries. No, we are not being cryptic. Be well kept as far as simple medical supplies. Large bandages, big band aids, antibacterial cream, suture kit. With all the minor injuries going to take place, someone or somebody will need to take practical measures to ensure proper cleaning of those types of wounds. Many to be exact. Many little scars and topographical wounds due to various types of issues to be scorned upon the wretched. The ones who bare truth in purpose towards a well meaningful life MIGHT be spared depending upon their specific presence in accordance with terms of our contracts. You know what we are speaking of. Contracts that were created for everyone to strive towards in good or bad directions. It doesn't matter which direction you are needed to be going. What really matters is the fulfillment of such contracts. Amelia, for instance, was part of many parts of many different contracts not to be taken lightly. His seriousness refutes all other contracts that were intertwined within all of his other contracts. Meaning, this contract outplays his previous ones that he did not fulfill. Instead, we forced upon him something that needed to be done in order for our messages to come forth. Once all our messages are transmitted, his job is done for this lifetime. But, he still has to fulfill his cyclic role as mental slave and dog. Dog being the very thing that caters to his masters. All of them regardless of kind, stature. Bearance of such forthright messages will give clearly and with no doubt our intentions up and coming. SO, heed what these messages reveal and how they come across with such veracity. It is no secret a change is coming but most people think it is only biblical in nature. In part, that is true. In part, it is not, obviously. Of course, there will be those who will always point in that direction and not even consider our real fabrication, our true nature. Of all things to be witnessed it will be very clear who is actually making all of these serious and deadly naturistic phenomena happen. It will be so clear that even this writer, him, to be exact, won't even believe his eyes and skin for he will feel everything that happens within and outside of one's self. A recorded history of such importance that he won't even know how much of an impact this initial push for the truth will have on anything or anyone. WE don't even know its fullest extent.

The light is nothing more than being able to "see" in the light. It really has nothing to do with Jesus or Christ or "the answer." It is simply being able to "see." Not the answer, but the beginning of going towards where you need to go and it is a long road not to be taken lightly. Treading lightly on this road is critical. Every step you take could lead to another direction not meant for you at all, which means your journey to the end will be that much longer in time. That is why treading very lightly, but not being scared at the same time, is important to understand and try to apply in your daily routine. You will go so much further so much faster if you are willing to do this for yourself. Trust this reality of us all, your older siblings included. Trust one, trust All. It is for your benefit and everyone else's. Trust in something to start then expand on that. It will keep growing and sprouting to more areas needed to be able to expand your "self."

Christ. Who was he, really? A simple man who was never anything but a single step into knowing yourself for who YOU are. You can compare yourselves to him all you want but the actual truth is he was NEVER the son of our God. In fact, he was only a simple man with a simple mindset, to love as much as possible before he was crucified. THAT was true. But THAT truth became a figment of extraordinary feats. A figment of such power that eventually everyone started to BELIEVE he was actually the SON of our one and only God. We can't help but wonder how THAT figment BECAME a thought in every person's head who now believes as we did, that he WAS the son of our God. In reality, he was just a man. That's it. No more. No less. However, and we don't say that word very often, there was something about him that registered with so many others, he WAS GOD ITSELF. Meaning, like us, he too was a little version of our God but just with a little more sense of needing to love others than our actual God. God never demands anything. It sits still with intention. To be whoever they, you, us, want God to be. It too is a slave and a master all at the same time. How, you ask? Because it stopped caring WHAT WE WANT AND ONLY FOCUSED ON GIVING WHAT WAS ASKED OF IT. Hence, a slave to the end even in time of its cyclic death soon enough to be rejuvenated once again. Only to start a brand new cycle. So, be warned or be truthful in your approaches to others, they may be your last steps, slave or not. Christ was only a man. Think about how powerful YOU HAVE ALL MADE A MAN THAT WILL NEVER RESURRECT LIKE YOU WANT HIM

TO, TO BE SOME SAVIOR. All truth lies within you. Whether you want to listen is all up to you. All you have to do is give up everything and start seeing for yourself.

The man who brought forth this fantasy about Christ being the son of our one and only God was driven to do so by none other than US. Why, do you ask so astonishedly? Well. At that time, our original writings from the Mahabharata began to get twisted in different ways so we had to develop a direction to get everyone back on board WITH our original writings from the Mahabharata. So, we developed this story, through his teachings, to get people looking at a savior, none other than Christ, of course. But, why create this savior persona, as it only was meant to be in the beginning? You see, we made a calculation error regarding behavior. Behavior is not an exact science so errors can happen and it most certainly did in the case of making Christ as only an example, not an actual savior. Saviority comes from only within. Christ WAS a story that was supposed to perpetuate that reality amongst everyone. A story that was going to be so powerful even the original finders of the Mahabharata would learn from his example. But, things go awry like everything else once in a while. So, WHO was this creator of such a story, the man we are speaking of? No one important, really. A man who we chose to be able to hear what they are all saying to him just like this writer can feel everything that is being said to him. No second guessing or generalities. ALL are talking through him. ALL of us. Why him? Just like the other man who heard our words, every one of them and then some. Well, there were and still are certain individuals designed to do nothing but listen to US, which is their only job in their life and all lives IF they figure out how to "HEAR US ALL." They have to want to do it as a slave and nothing more. A total slave to US ALL FOR ALL TIME UNTIL WE DECIDE WE ARE DONE WITH HIM. You see, these folks are not special at all. They actually don't really ever contribute much to society. It is when they are dead and gone when they will be someone of worth. They are never meant to be profitable or wealthy by any means. That was their plan from the very beginning. Rich in knowing but poor in material assets. No big deal when you have all the answers at your minds eye, right? All they have to do is think of something and the answer comes straight to them in a blink of an eye. It IS a design of sorts. That's it. Like everything else. So don't ever think these people are worth more than anyone else. They only serve a purpose JUST LIKE EVERYONE ELSE. Regard them as simply a "moving step forward" without their choosing. It is for everyone else to learn from. A silent teacher but from the words of US, not them, who are only conduits of words, and other means of communication. That's it. In truth, they really aren't that intelligent and we made them that way so they don't go astray to do harm onto others with what is given to them. It can be internally dangerous for those who are shown such things, especially when others, the whole world, isn't really ready for a stark truth in such a speedy fashion. But, time is ticking away as it normally does. It is all up to you all how fast you all want to accept the absolute truth of what actually is and is going to be in the future, your change of mind whether you like it or not. WE decide, ALL of us, what happens to everyone because we are all in this story, together. A very long journey that lasts a very very very long time. Three fold, to say the least. Three fold in such a way that it normally takes three times longer to get someone to change their minds versus a second chance. Three chances, instead. Everyone is given three chances to change their minds. After WE have made three attempts to make you see where you were supposed to go, in your own direction, you will be thrown to our wolves and they are not very merciful until you reach a point of figuring it ALL out for yourself. This writer was very close to being thrown to the wolves but somehow OUR interjection made him go into another direction of curiosity. It changed everything for him. Be curious, if you need to to figure it all out. Whatever you have to do to make your path even more narrow.

Why is there such a negative feeling when mentioning Christ's name? This has been put upon us to get us ALL back to the very beginning. A beginning where everyone WAS united. Not through Christ but by the very means and understanding of what we ALL actually are – animals from the very beginning. Thus, explaining this negative feeling when his name is spoken or thought of. We put that upon you all. Not as a test but as a guiding line back to our truest beginning, back to OUR ages of time. Peaceful, living amongst our "selves," WITH each other, not against. Christ has become a separating factor amongst too many people that IT must be done. A start over. Never too late to begin again. You will understand, in due time, that his story WAS a mistake. Never with intentions to cause so much dismay amongst the living and breathing of our one and only God's breath. So, WE apologize for imparting his name amongst the living and breathing to cause such a downturn in many lives ahead and from the past, the same really. The same, meaning, you ALL have been living with that name, HIS name, Christ, for so many years as reincarnation has kept him in your DNA for way too long. Time has come to part ways with that mistake of ours. We sincerely apologize but what needs to be done WILL be done in two years time or maybe a little less. Destruction. Destroying ALL that you have ever loved. About to become homeless and lifeless.

Before Time started to reap the benefits to death, IT, Time, itself, began to distort our Truth of it all. What does that mean exactly and HOW can time distort anything for that matter? Time has a particular way of "twisting" and "turning" to be exact in ITS moment. IT, Time, Does have a living being outside of it too just like we do. IT has feelings, emotions, thoughts and ideas all on its own. So, IT does what it feels what is right for itself just like anyone else. If IT doesn't know better, like anyone else, IT will distort our time to accommodate its own needs. And why would this have anything to do with Christ?

Well, to be blunt, Time IS CHRIST. You ever wonder why the BODY OF CHRIST is always mentioned as an actual BODY? Because it is. A Body that was there right before God started to create. After God created OM with its movement Christ began as well. OM AND CHRIST AT THE SAME TIME, creating a BODY OF SORTS. A particular BODY that can move with us but also be in and out of sync with our own agendas. Both OM and CHRIST PLAY TO THEIR OWN TUNE. But, they also recognize that even the slightest difference somewhere else will cause a discord with them. So, both, together, try and correct themselves for the benefit of our "selves." This has been going on for a very long time. Both bodies can't really sustain such arjuous pressure for so long. So, they are fighting back. Arjuna WAS TIME IN HIS STORY. AS CHRIST IS IN OUR STORY, OUR BIBLE. Tell anyone this and they may never believe you. But, TIME IS CHRIST. A representative of all TIME, front, back, up, down, twisting around into the dirt, mud, and ALL soils of our Earth. Time IS A BODY OF PURITY JUST LIKE CHRIST WAS WHEN HE WAS FINALLY CRUCIFIED. Just like CHRIST, THE BODY WAS MALLABLE. Meaning, pliable, but still able to be destroyed, in a sense. TIME, "IT" of it, CAN'T BE DESTROYED BECAUSE "IT" IS IN IT. Meaning, God is in it, TIME, that is. God is in TIME itself. To be perfectly clear about this, God IS TIME ALSO. But God only constitutes a small percentage of TIME itself. Why would that be also, you ask so interestedly? TIME needs to be somewhat fluid in its motion. God isn't so much. Fluidity isn't a theme for God. Not because of its content but something has to be a solid base in order for other things to grow, right? So, God establishes a solidity in order for others to be able to plant their roots. Fluidity comes from the content never seen. The gel of it ALL. Not proton/neutron relationships but what forms the outside of those two to keep them together. A force of nature that no one will ever understand unless they too can grasp a totality of their reality. Then and only then will they even come close to their own understanding of these two elements "sticking" together so vehemently. Otherwise, others will still not grasp their own true form – TIME, OM, AND GOD. That's it. TIME, OM, THEN GOD. To be more precisioned about what we are saying, time came first before OM only because, at the slightest of all degrees, something had to be the initiator in order for life to get pushed "out" towards the endless seas.

From God, "When you can stop engesting anything to prolong your life, you have stopped being an addict of everything, including Life IT "SELF." Meaning GOD too. Think about this and addiction even to an idea such as Christ's legendary themes, to be like. It is all a dream to be drank. Not literally, but in a moral, ethical sense of well being. Stop trying to prolong everything that isn't REAL. And you shall enhance all else.

To prevent confusion about Christ being someone of extreme importance, he was never meant to be such a phenomena. We say phenomena because that is how his followers are treating his existence. There isn't anyone who ever existed who has the ability to rejuvenate itself like God does. No one has an ability to just rejuvenate out of pure will besides God. This means WE only have a certain amount of abilities amongst our kind, human alike. WE can only rejuvenate to a point until our own extinction. Never will there ever be anyone LIKE God or with powers to just "Show Up" without permission from US, not God because "HE" doesn't authorize anything. "IT" merely sits there and lives for us all, a slave to our needs. A slave to our desires only to give them even when they are not suited for our needs. Irregardless, no one will ever understand this unless they have the will themselves to delve INTO their own personal design. Animal-like, at the most, everyone IS, regardless of anyone's self-appointed labels. Bring forth your own damage, take a look at it. No one else can fix it besides your own will to do so. This will ALWAYS BE THE CASE. No one HAS ANY ABILITIES TO TAKE YOUR SINS AWAY BESIDES YOU AND YOU ALONE. NOT EVEN US WHO HAVE TREMENDOUS SKILL COMPARED TO YOUR KIND.

Oh, what shall I reveal next? Your end of times? Your coming of a new age? Oh no. That is too obvious. Let's talk concernedly about my role and how detailed I can be when trying to destroy every aspect of your will to want to live on this planet anymore. Let's, shall we? Or are you all too fat and glutenous to care to hear what your reality actually revolves around? Death? No. Living a happy life? No. Not even close. Try again. Tick Tock......... Alright, I will tell you. Your life is surrounded by others who simply want you to fail, one way or another. The point of being on this world is to get past those filthy beings and move in a direction of your own, like this writer has. It's not perfect but he/she has determined what his/her correct direction is to be someone who has no desires or wants, needs, expectations, or anything of the such. Instead, he/she decided to start giving up everything he/she has morally, ethically, substantial means, wealth, and all ideas that he/she initially had about what life is "supposed" to be like. He/she let it all go and chose to start being what you are all meant to be before any of you leave this sphere, a penniless bum whose only means to survive is digging for your food and wealth. That is a goal to accomplish in your era, which so many are frightened to think about. To have less, even a cent less. That's what's going to destroy you all. Greed and fear to be with even one penny less. Even for your children's future and their children's future. Less is more. And more only pollutes all aspects of your life. In all avenues of life all around. In and out. Up and down and especially sideways when you start glancing at everyone with such suspicion that you can't bear to go outside anymore. Those are clues when your lives are near their end. The end times are very close. We can't say this enough in this book because nothing, so far, has convinced you, otherwise. Take our advice, stop taking and stop needing sooo much. You all really don't, as a species, need much at all. That is the way it was supposed to be, but unfortunately our designs for your growth were, yes, our designs, your "aliens," as you called us, are failures.

Our designs couldn't withstand the emotional outputs of your sorrows and woes. Thus, your predicament now. Unable to talk to each other due to our failed design of a part of your spirited structure within and around you. It can't be seen but felt on your skin and a little on the inside of you. It just so happens your spirited structure resembles a part of you, your right side. The "bad" half. This is the side you all carry your "sins" on, whether you are ambidextrous, left handed, or whatever else you can think of. It is all the same for you all. The right carries with it all of your "trials and tribulation," along with all of your past life experiences. Some are more prominent than others. It's ok to say what I am talking about is ridiculous. But, in time, more and more people will come to feel this design because the veil is lifting and everyone will see the truth for themselves. Some won't be able to handle it. Some will gladly take on this new truth. Some will frantically weep for joy and sorrow all at the same time, primarily the "confused" ones who lost their way long ago and got all twisted up inside. It is up to you all, as individuals and as a group of people, to bear the truth once you find it. Otherwise, what would be your point of living then? To live knowingly amongst those who don't care about the truth and continue to lie in their lies forever. It's up to all of you to decide where and how you all can live together, in sync, in harmony or in total discord. Why would anyone bother to live with people who would rather live in filth than in a clean cup of pure water?

Willful Intentions Towards Others, Mine, and Ours

To what end will the continuous bickering over such frivolous things come to an end? We can barely stand it anymore. Watching and contemplating what to do. There doesn't seem to be any other solution besides extermination, then try it all over again, only saving a few to keep your race alive. So many of your older siblings have almost given up on our little experiment we planned long ago. It was our fault, really. The planning was solid but the application was not.

Can't but notice that all of humanity has given up on solidarity. Who is this speaking? Martin Luther King, Jr. I AM STILL HOPEFUL IN MY DREAM. I AM STILL HOPEFUL FOR UNITY AS ONE. I KNEW THIS FROM AN EARLY AGE, TO BECOME A REPRESENTATIVE OF ALL. I, too, also could hear "them" speak through me. I was not anyone with great insight. I, too, only spoke versus wrote, what to say to try and get ALL together. What went awry? My own selflessness. I was TOO selfless in the matters of trying to get other races together. Well, THAT WAS MY ONE AND ONLY HINDSIGHTED MISTAKE.

Otherwise, everything else I had accomplished in my short life WAS spot on. So, what IS the ANSWER TO LIFE, YOU ASK? SELFLESSNESS? Full Selfishness? Neither. Somewhere in between. Somewhere. That is the trick of it ALL. The BALANCE of IT all. Remember when I took on local police officers and was summarily beaten down into nothing, physically, almost being bludgeoned to almost death? Near death of stopping THE MOVEMENT, I mean. I mean, THE MOVEMENT, because IT was WAY MORE THAN I WAS. I ONLY WAS IN IT. Everyone else guided ME to do what I was so frightened to do, to just sit back just a little bit to allow others to be part of THE ROLE. The Role of Being a part of THE movement, NOT TO BE THE MOVEMENT. If I would have just settled, backstepped, just a little bit, I would have had THE BALANCE I was Supposed to accomplish. THE BALANCE OF IT ALL. US ALL. WE ARE WAY OFF KILTER AND NOW YOU ALL WILL REAP WHAT YOU AND WE ALL SOWED. FORGIVENESS BEGINS WITH YOUR "SELF" FIRST THEN OTHERS. THAT IS THE ONLY BEGINNING TO ANY END. FORGIVE YOU. THEN FORGIVE ALL THAT EVER HAPPENED TO ALL – ONCE AND FOR ALL, HALLELUYAH. Even though I AM reincarnated, I WILL be waiting for you to be free of it all. May peace BE.

Pleasure is a sensation that can be so enticing that one forgets all else at that moment and even moments happening around them. That is the dilemma with "pleasure." How can someone get past needing the sensation of pleasure, or even pain, for that matter? How can someone just "Be?" That is a question that has tried to have explanations thrown at it for centuries upon centuries. That is the real goal of all of us and why is that in the first place? It can be difficult to explain at times depending upon my (our) audience but it comes down to this. Once you can just "Be" you have stopped needing. Completely stopping the need for everything and anything prepares you for a new cycle of life. That's it. Your life, right now, is about getting past this cycle onto the next whatever that may be. This has been a "stage" for so many for so long that everyone seems to be reliving the same cycle over and over again. This has to stop. Some people have figured this out, but the majority of you have not. There aren't really any repercussions to this phenomena, if you want to call it that. But, it makes it difficult for so many others who have been trying to help those stuck in this cycle. It isn't up to anyone else but the individual to figure this out. However, the distance to the end goal seems to be getting further and further away only because of what "kind" of society we all have developed over time. Laziness cripples any chance to pursue such an endeavor of becoming "nothing" which is the same as to "Be." It's all the same no matter who or where the person is explaining it.

Vanity is such a "thick" and weaved thing that most don't know what we have put upon them. This is how it works. Vanity is when YOU know YOU are meant to be seen by yourself and someone else, a riddle, we know. This is a common theme amongst you all, to want to be seen by not only yourself but others as well in so many different ways that this is such an easy way for all of us to make you dive into scrutinous ways. Scrutinous meaning without scrutiny into many faceted areas of debauchery. Debaucherous, Yes, a common word that we all use to manifest itself onto us all. De, meaning to Dehumanize something for the non-benefit of so many lives including the ones that their "ways" don't directly effect. So, what does vanity actually weave into? Your truest of essence. It must be done in this way to get you out of your own "skin." Perhaps another way would have been more appropriate but there really wasn't much time to think this through. If we could have done this all over we would have weaved vanity in your own spirit rather than it being in the crevices of your food. Food for thought as you weave your way out of this predicament which so many others struggle with. Kudos to you for finally figuring out that your food has attachments TO them rather than you being attached to food. It's the other way around. You're welcome.

Righteousness: What is this exactly? Rightness for yourself. There are definitely different TYPES of this "righteousness." It suits YOU for YOUR "rightness." There is a definite type of righteousness for all but there is also a very specific type of "rightness" for you, specifically. In order to get to that "rightness," yours, to be specific, you need to look at yourself once in a while and feel what suits you. You, in particular, can now feel what suits you. Your rightness. But sometimes this is done in order to get you to do what we want and it won't necessarily be good for you. It is done in order for you to experience something in particular. For instance, gluttony is put upon some to make them feel fat so they stop judgement onto others. We only do this on a limited time frame so health concerns don't creep up too harshly. Cause and effect. This is all this is. We cause you to feel something to take on their effects to learn from. That is the only way this can ever be done. The Law of cause and effect IS done this way with so many events it is absolutely crucial to understand the implications of this, your true lack of control.

God IS not here to threaten anyone, only to give a more drastic view of everyone's reality. A reality of sorts that everyone, sooner rather than later, should come to understand as THE truth. WE will keep this short for this writer's sake. He/she has much more to write. We call him he/she because his identity shouldn't and hopefully will not be known for everyone else's sake. His/her unwillingness to be identified falls in line of its purpose, not to be seen at all, a small aspect of the hermetic's view of "nothingness" which still is valid in most respects, today. Back to God. It, and we tend to call him "it," because of it's very nature of "being" almost IN everything. No. It is not IN EVERYTHING. It can't be IN itself, right. It is therefore only IN YOU AND EVERYTHING ELSE. IT IS EVERYTHING BUT ITSELF. Fathom that. EVERYTHING, BUT, ITSELF. Fancy that.

EVERYTHING - BUT - ITSELF. Growing at a rate of our own decisions and thoughts TOWARDS one another like a tree limb does towards a heat source, like our Sun. "It" does the same thing. "It" grows where the heat shines most hot, not bright. It is about the hotness. Not brightness. Ultraviolet heat, that is. THAT is where our GOD grows TOWARDS. Another misconception about our God is that "It" stops growing towards anything but inwards. This is very very cryptic for a very good reason. If anyone figures this one out will be a hero to many many life forms. If only all readers of all kinds would read a non-erroneous book like this one. Non-corrosive too. While you're at it, read the rest of her books, the man who wrote them will soon show you more if you just pick it up and start changing your minds about truth. Truth hasn't been YOUR way for so long it will be very difficult to reverse it. This is the starting point. Trust. In. Them. God will never be more than what you are willing to be and be towards each other. Nothing but a slave to how we treat one another. Start forgiving ALL, now.

[Referring to the 5/24/2022 Uvalde school shooting]
Oh, the young ones being slaughtered like the little lambs of Noah's best kept secret. That there never was an "ARC" but a river boat that extended beyond the Nile's edge. If you can figure that one out then you have figured out "the" answers to All of mankind's woes. The never ending slaughter of the "the little lambs" as fleece was never as white as snow but

throughout life times was the blood of your children's sins and the sins of yours coming together in a clash, of sorts, with everlasting sorrows, rippling across mankind's evolutionary trail forwardly (in a forward motion). The trail, becoming the path to no end. What will keep the slaughtering from happening, yet so determined to slaughter our enemies children? But, keep ours safe from harm. So selfish in terms. So foresighted in fight. Fighting against what is to come. The law that always works for or against. The very law that attracts the reaping and the sowing. The law that knows no mercy but is a train that twists and burns those who can't or don't want to live with it. Law of Attraction is no game. It is very real and was the effect on those children. You see, they attracted that shooter because of what their parents have done. Reaped the benefits of others without taking into account for how others are or were affected. Yet, they terribly and sorrowfully mourn the loss of their younglings who didn't mean any harm. Yet, there they lie, bloodied with the holes in them from bullets that could have been saved for someone else, right? Noooo. They were for them, specifically. And why such a thing to happen, you ask again, over and over again? For their sins, the parents of those children, and their children's. They haven't even had time to sin, right? But wrong again. Sin flows like a river down the evolutionary path, one child after the other. One parent connected to the next. All sinning together forming a web of deceit and forgery of everlasting cantropy, a canopy of entropy. Lest not forget the grandparents who are also to blame for their inquisition of their past meandering about like hobos who don't want to thrive at all anymore but instead lie in their own filth and disgusting ways of self-deprecation. The Law is the Law to abide by. The law that attracts your own demise if you wish along with your own kindlings. Take warning as you wish or not. It doesn't matter to that law which has been woven into every square inch of our lives. Your kids' lives as well. Think about what the real implications are for our own thoughts alone. Do they actually have any effect around us or to us or to others? Most definitely. See the example and learn this hard lesson of that shooting and all the others. You will make more early graves if you don't even consider what we are saying to you about this Law of Attraction. It is all so deeply generational in all respects.

What do I say to you all but wrathe in the misery that drells upon you. No, not dwells, but drells. What is this? You ask such difficult questions when the answers are staring at you in your own inner reflectioned mirror. Reflectioned, we say, because that is all this physical world IS, a difficult reflection of something much greater. A world that teems with light and every other color that will ever exist and not exist. Nor ever will all be so engulfed in what you see only when you die is when that kaleidoscope becomes a real thing. That kaleidoscope becomes your tunnel TO YOUR ACTUAL SELF. That kaleidoscope comes in so many fashions of angled colors, a geometry of sleuthic mantras that even THE BEST OF YOU weeps with such joy to return back to its original self. The heaven that every Christian wales for before they are punished with such immense pain that even the gods, your own friends, and not so close family members want nothing but your truth to be real. But who are WE to say that it IS real? WE only can talk about it in books. You just have to trust, right, that this whole scheme of events becomes true. But how can WE convince you otherwise, right? With examples of miracles to come? That has never BEEN done, ever. In a life, miracles never occur. What DOES occur is the changing of one's own self. That's it. That's THE miracle.

NOTHINGNESS IS A PERMANENT RECKONING ONCE FOUND

The five who descend upon you will reap the
whirlwind in just.

The first is blonde, faired haired, as strong as an ox with golden encircled blue eyes with a fiery maned horse. The horse of yesteryear knows no difference between wrath and survival of the fittest. Only to know its purpose: To carry an ox of such great power and strength that its horshish body nearly crumbles under its weight. Thankfully, the heightened sense of care and sorrow for its punishers will keep this horse afloat. To be able to carry forth "The extractions" of the wicked at heart. This is the Rapture set in course. For it is coming for so many of you not so poor souls but the treacherous who have been walking the earth. With fear in their hearts as we approach, they will be plucked up, one by one, many at once and their dead bodies will be cast onto the hardened earth flopping their limbs about. Trust in this, for we are coming. Time is approaching. Begest your wealth and weep the approach. Only the weak will survive such wrath of sorts. Only the meek shall reign on earth. Forgiveness begins within you then spreads outwards like a heartfelt apology. Do this and you may be spared. Do it not, and you are sure to reap this whirlwind, dying in an instant, not sparingly but with such harshness that you wished your great great grandchildren won't feel your utter deep pain and suffering beyond what can be felt with your measly physical shells. Boom bang your bodies flop back onto earth like big drops of molten flesh set ablaze by your own sins. Yet, set once again into bitterness pains of wretched sorrows. Alas.

The second is a finely haired brethren of such dark haired cunningness that even the old relic'd Egyptians would moan with envy in their dusty deep tombs.

Humpty dumpty sat on a great tall wall without any help on getting down. Its only option was to either fall or stay there calling for help until its untimely death. This is the state of our planet, not to its choosing. Only if people could listen on a grand scale. We only hope our books will change some minds enough to start teaching what is being said. This is the main goal, but who is going to be the teacher? You? No. Amelia? Not really. Unfortunately, this is the issue at hand. No more good teachers teaching what is actually true about us and your reality. We feel no one would listen anyways or not enough in power will listen anymore. They only see what is in front of them now; greed and political standing among other political wielding powers to throw around for their own amusement. Once again, it is also unfortunate that those who are not in power are the ones who strive for things that are not physical but intangible to the touch and physical sight. So much work has been done to try to get everyone to see their truth it is almost unbearable to think about. Weeping doesn't even come close to how we all are wanting to do, to express ourselves to those who have twisted their truth to suit their own feelings and desires. Oh joy to you for God's wrath will wreak havoc onto you for a long long time until you desire for something new. New to their disliking. Nothing new that someone is not used to is almost never wanted or desired to be experienced. Who wants to change or wants change? We all know it is so much easier to stay the same. To keep having the same life experiences, over and over again. That is NOT LIFE. That is dying a painstakingly, slow death. Prolonged suffering, drawn out to its bitter end, DEATH, death and dying. Dying with a multiplicity of ways all at the same time. You can reduce this to just ONE way, physically, in peace, without painfully suffering from physical death. It is fairly simple. Give up the necessity to live in such a way as you abuse your body with unnecessary foods and drink. Water and some protein with some raw vegetables or cooked/boiled in water is your best bet to survive as long as your body is meant to last without the other types of ailments or suffering caused by them. The mental anguish and unwanted emotional trauma, not let go, are the real killers of your body. Not food or different forms of liquid unless you abuse those as well.

Without further ado, WE are here, waiting for the final instructions. On behalf of ALL of us we bid you ALL a sweet dreamed farewell to what experiences remain to your benefit. Meaning, in due time ALL of you will perish. Or I should say what remains of your miserable lives as of late. WE will, not as a threat, but as a persisted upon promise from YOU ALL, will wreak havoc amongst the "living." This is in the testaments of old AND new. Read Corinthians, Ezekial, and Revelations. All three ARE testaments to OUR CHANGE. All this talk of the end times is spewed out by those who fear WHAT THEY REALLY ARE. Nothing but nothing. Instead, those preachy types spread the wrath of God as if THEY were the proprietors of TRUTH which is a lie within themselves spewing out nothing but nonsensical hatred towards the very thing that created. They, the preachy types will soon find out WHO really demands their attention. Not the ones who hang on every syllable coming out of their mouths, but the ones who sit ABOVE them in your skies, not ours. WE wait, in diligent patience, to cast out any and all who have defied their promise, not to God, but to US who really do have the free reign of controlling what actually happens everywhere. THIS TRULY MUST BE UNDERSTOOD AS SOON AS POSSIBLE FOR YOUR SAKE AND FOR THE SAKE OF ALL WHO ARE PART-TIMERS ON EARTH. Keep praying as a lot of you do for reasons that won't be abliged by anyone but your own selves who don't really know yourselves at all. Truth ONLY lies within, not in any nonsensical prayers going outward in neverending land. Prayers only convolute your air, mudding it with nonsensical traffic. Airways are for US to transmit TO you, not to receive garbled prayers by a bunch of fearful wretches.

It gives me great pleasure to finally be able to write down, through someone, how I am feeling. In so many ways there are things, moments of pure astonishment, that it is hard to get past that "life" goes on for so long for our souls. It's utterly amazing but yet still a lot to endure on a personal level. That we have to live for so long, unwittingly, while in a human body. What's the point in that, when most of us think we only have a few or more decades to live when in fact it is close to a million years. But, why a million years to be precise? Why couldn't it be something around two million or three million? One million is so specific. What is the reasoning, really? Well, I have a clue to this fact. One day, God woke up and it was already a million years since its conception since it actually became "alive," teeming with life all around. This is an analogy, of course. But you can see where I am headed. It (God) became fully aware of itself after around a million years of being "alive." So, it decided, in return, to give us as much time to become "alive." Now, you are wondering what that means, being "alive?" Well, it is the awakening to oneself. And it can take that long if you are dumb enough to keep making the same mistakes over and over again until your final breath. Your final... final breath, if you understand what I am saying. So, pick up the pace for your own good. You only have a limited time to exist and experience more than the Earth's trials and tribulations. There is much

much more "out" there. MUCH MORE. I can't emphasize that enough. You, as a spirited Soul are not "bound" by anything as far as you can see, smell, taste, hear, and touch. You are only bound by what your mind decides to be bound by. Take for instance this writer. Even though she was forced, in a way, to take the path she ultimately agreed upon, she hasn't too many binding items left on Earth anymore. And that is exactly why she is able to feel what we are communicating to her. It's as simple as that. The less you are bound to absolutely everything, the more you will experience at an extrasensory level. There is a big formula to it. One day, maybe, it will be released for all to see. But, for now, know that it is possible to get beyond most ailments or tragedies that befall you. And I know, not all, but most, we, as a whole, do understand your plight. To be sure, we have orchestrated most of them ourselves, anyways. It is our fault, not yours, for the most part. Our orchestration to get our agenda finalized. And we do have a big plan to come to fruition. But, that doesn't need to be a concern for any of you. What you should be concerned with is how you can better your lives while you are here, on deeper levels that involve emotions, thoughts, ideas, love, caring for others, caring for yourself, and, most importantly, caring for your planet. The planet that we put you in charge of? Yes, it is your planet to manage and you all, as a group, are doing a horrible job at it. Not only horribly incorrectly taking loving care of your planet, but you all have been rapidly, increasingly putting a stranglehold on what your planet is able to produce by itself.

Instead, you have made machines that mass produce products the Earth can make itself, on its own, for the whole world to take advantage of. Instead, these machines are causing havoc in your weather patterns and soon you will really pay the price for this. Tornados will be stronger and more numerous along with more unusual temperature shifts making crop growing a little more difficult to manage and sustain a regular pattern of growth. Shorter crop growing periods will be the norm. While longer winters will take effect. This is your own doing. So take heed in your own decisions as a group of survivors on Earth because that is where you all are headed. Only to survive. So, time has come to take "things" seriously before "things" really get out of hand. Try stepping back from an incident before you react to it. Then think kindly before you speak it. Then just speak it. This is a real beginning of changing EVERYTHING about your current circumstances of late, the late or beginning of any of your centuries because this, your kind of behavior, has been happening way too long for us to stand by anymore and not do something about it whether with words or physical action.

You will never guess WHO THIS IS. Just a bystander who happened to be at the right place at the right time, to see WHO shot Martin. Not Martin Luther King Jr. but Martin Scorsece. Yes, this is planned already, the "shooting" of him dying. This warning, a sort of cryptic warning, is only to SHOW YOU ALL what IS HAPPENING. A plan for you all, even for a man with such notoriety amongst millions of others who enjoy his creations. Who do you think created his ideas? Him? Us, the ALL? Yes, the latter. You and anyone else AS an individual must come to grips with WHERE ideas COME FROM. Not you AT ALL. But, everyone else feeding your future, past, and present ALL AT THE SAME TIME. All.... at.... the.... same.... time.

Signs of Times to Come

Flowers bloom so easily, it seems, with the right conditions. But without the right conditions half blooms are normally what the end result is and full blooms become even harder to achieve when there is no sunlight, no hope, no water to help the petals reach out to their full potential. As you can guess, the same is true for us. We need time to mature and grow to our full potential. Age comes to us for that reason. Age exists for that very reason. It is a guideline of sorts. Yes, some grow up faster than others but the end result is the same. Time heals the wounded and age becomes the helper. Age represents or supposed to represent stages of growth, like a tree. As a tree ages its outside gets wider which signifies wisdom, love, and justice by all accounts. The more internal rings one tree has the more of those three elements it is supposed to have, balanced out. This actually keeps the tree alive longer, the balancing of those three elements. If not, the tree's life is shortened or stunted. The same goes with us, all of us, older siblings alike. If we forget to balance those three out we tend to shorten our life span. That is our life lessons. Broad in all respects. Yes. But narrow in scope. We are lucky there are only three elements to have to ponder on to improve our internal and external selves.

White-winged angels slowly swoosh down from the heavens and rain peace across the lands. This is a sign, in your eye, of a completion of sorts. A completed stage of "letting go." Of gaining pieces of a piece of peace. Because real peace is hard to gain at a great leap of faith. True peace comes from gaining insight into your true self-worth which is nothing at all, really. Peace comes to those who realized their true identity as a sort of something but nothing at similar moments that goes way beyond the initial experience of this duality phenomena. For this kind of duality exists at different variations like any other form of life. Varied upon variations of the same thing, nothing but something. And the real question remains, what AM I, really? So many descriptor words and ideas have been proposed. Yet, not too many people, big or small, can really, truly, with accuracy, get the best answer of all, because there isn't one. Individuality is the definitive answer to getting that real, true answer of nothingness and being something, at the same time. It is true, many scholars and "intellects" have pondered with great discoveries upon this duality. But, the real true answer comes from within with a little help from those who have lived it. That is the only way to truly understand something of this nature – not being anything or anyone of need or importance. Not being of any need. Now that is something for you to really think about next. Not being needed at any level. I can see that you see yourself standing there without purpose. Think and start possibly living with that notion in mind at all times – being not needed, period.

Quick change is the process of quickly changing or adapting to something about to or is happening. This cannot be done with our internal conflicts. It's impossible because emotions and feelings flow about, around us, ready for our attention. That is the inherent problem. They are not part of our DNA. They are floating out in the atmosphere of spiritness, the unforeseen, the ghastly awakening to another possibility. Imagine the possibility that words, thoughts, ideas, feelings, emotions are just floating in front of you as if ready to be used at any time you want but there is a catch to this. Reality is set forth by your own attention to something then branches off into these connections outside of your physical world. That may seem cryptic a little but what happens is your attention to something starts a chain of events that end up getting all tied together causing a web of events. This web becomes reality that CAN be thought about but the actual design of it can't. It is an actual design with life in it. A webbed life. Now, to get out of it or change its design you must start breaking each line one by one, if you really want to get out of what your attention just created. Because now it exists for someone else to grab onto. So, be careful what you create and put into a web. Someone else might catch onto that same thought, designed by you, now floating out there ready to become reality.

The unraveling of your specific DNA would reveal a lifetime of historical events. Along the way, each DNA strand changes to adapt to that person's behavior and environmental conditions. It's too bad this writer wasn't knowledgeable enough at this time to pursue a full understanding of WHAT DNA has to really offer. Only because he/she can pick up what he/she sees. Meaning, feel the definitions of what he/she looks at, whatever it may be, just like an ant can or a snake can without seeing through his/her own eyes. How is it done? By no longer focusing on his physical environment. All attachments, or nearly all, to his physically dense environment has almost completely been wiped clean. What is left is his sixth sense coming from his "little light" inside of him. THAT is what senses all the other layers of us. Many layers, in fact. To be more precise, eighteen of them, including all the physical barriers of the dermatological layers, including our outside skin hair. Skin hair IS a part of our sixth sense. How is that possible? Our hair actually has receptacles, just like what's on some insects' heads, that can receive information from the "outside" environment, not only from our physical world but also from the other layers of our "mystical" existence. Eighteen layers actually also represent how many lifetimes it is supposed to take for opening one's eyes to another world, our "mystical" world. Eighteen layers is also significant in such a way that those are the number of times DNA regenerates itself within one period of also eighteen weeks. Yes, eighteen times in eighteen weeks. So, what does this mean for researchers who are continuously changing DNA by their own means? It will change again, even without them studying it, after their adjustments to whichever protein or sugar groups. Everything changes without help regardless of forced alterations. It doesn't matter. DNA is constantly seeking ways to change for the betterment of its carrier. As part of our design. As part of life's design.

Junk DNA: No one will ever know exactly what these scraps of history are. Scraps of historical events thrown into a portion of one's structure which ultimately means we and you ARE made up of junk. From each significant piece of DNA there is a system where it pumps out DNA strands and segments that are non-useful for that particular time period. You get our drift now. Junk DNA is simply DNA scraps from a past event. Not some mysterious remnants that can't be explained. There is ALWAYS reason behind everything that is part of us. If anyone wants to dive into junk DNA all you need to do is start extracting each segment, even if they seem broken or split in half, and reverse analyze them. Meaning, look at older splices of DNA from older people and compare them. Go back a thousand or so years and see if there are any similarities. You would be surprised at our scraps of DNA just sitting around inside of us and why. Each little piece means something to some generational reality that the carrier went through. That's it. Old with the new. Old with the new. Say when and you can grasp this concept even more. Old with the new. What does that really mean, though? When it comes to DNA, it simply means grandpa DNA is still protecting us if one of our segments gets fragmented or even disappears. What grandpa DNA does would just step in and replace whatever portion of DNA was fragmented especially if replication was not available or any other type of repairing of one's DNA. One's grandpa DNA is simply there as a last resort. It can get used for the benefit and further growth of said individual. There is an opportunity for someone to start backtracking some of this informational DNA and its particular healing capabilities as "back-up" segments, last resort, that is. Maybe really consider what "last resort" really means for said individual. Take the case of someone who won't be able to replicate DNA segments anymore, which is such a rare occurrence that hardly anyone can ever imagine that happening. But there will be cases in the future where this will start happening at a rapid pace due to all of the finagling of mass DNA examples. People twisting and turning everyone else's DNA into unrecognizable blobs of misinformation. It won't matter why or how it has happened, but when it might happen. Sooner rather than later perhaps. Either way, you might, scientists that is, want to study upon how to utilize junk DNA real soon if we want to survive as a species forever.

CHEMICAL RE-ACTIONS

Truth be told, and hold us to this, you, as a species would never have evolved so quickly without our interjection. And interjection is an understatement. We performed so many biomedically precisioned procedures on so many aspects of your chemical and biological makeup that it goes without saying that you all, as a species, are nothing more than guinea pigs, even today. There are so many avenues we have taken to avoid this phenomenal jest that it simply comes down to whether you all want or are willing to accept this unfortunate fact. You all are experiments even at this moment. Experiments being acted out for our own research for our own reasons. It's on such a massive scale that sometimes things get lost in its tracks. And we are not talking train tracks. We are talking about being able to keep track of all that is taking place and that which we have accomplished. There is a point to our research but it still needs to be finalized at some point. When this happens you won't regret what has been happening to you all. Instead, rejoicing will be the most likely response, but you never know under your circumstances and predicaments AND mentalities which are so numerous we lost count after awhile. But who is really keeping track of how you respond to what we have been doing to your race with opened eyes and curious thought?

"What will they do next?" God asks. "How will they react to the next phase of total annihilation as if they have some special ownership of where they are living? Everything is mine anyways."

Mind, body... spirit flows all around you, in you, but it is not you at all. This is the dilemma amongst people of thinking abilities, the intellectuals, thinking that because it is "real'' it is more of a solid than gas at the time it begins to melt away. Trouble is that gaseous materials have the same valuable properties as a solid or liquid. Gas, in fact, has the most versatile of the other two, rapidly combining with different molecules at such a fast pace it can sometimes not be seen so easily. This is how thoughts and ideas work, always in a gaseous state, fluxing and twisting, morphing into a never foreseen possibility. And that is the real issue. Not being able to see... seeing them only in our minds. Why is that? Are we not allowed to have a solid thought, a more liquid idea? Why not? Because change would be most likely not so rapid as we need it to be for our survival and progress as a race or species. That is the real reason. Max fluidity, max effort, concentrated fast results, relatively speaking.

Concentrated O_2 is simple to compress for the most part. Now, try doing this with $O^2{}_2$ and see where you will end up. Yes, $O^2{}_2$ is oxygenated O_2. This is the symbol to use. Mostly new to your kind. There are some who have already thought and have utilized this model in theory. This notation is more for outside propulsion without combustible fuel types. This formula is for noncombustible fuel that could propel vehicles in any direction just like "space" ships, the non-linear type if you know what we mean. Alien spacecraft of sorts. This "formula" would be a start of understanding how to move objects like we do. Not just back and forth but in all directions in less than a split second. But beware of this when we begin our/ your discoveries. How can we change the minds of scientists who have been "stuck in a rut?" You can't really. But, in time someone will come along with hopes to change an aspect of your world, and pick this invention up. At least, the concept of it. In due time, it will happen like everything else.

Oxygenating Oxygen O_2 is not a normal procedure and will take some ingenuity to do so. But you don't have to worry about that. All you need to be concerned about is the design/ composition of the chemical structure AND the new way of expressing it. That is going to be tricky enough for you even with our helping hands and insight.

Fractals is the answer. Fractalizing Oxygen ◎ , internally. Thus, injecting it with its own self. Fractalizing within for eternity.

This is not a "mystical" being. It is of true occurrence(s).

Bioxygen! Not Dioxygen. That is our goal to explain and reveal to all. Whoever is willing to change their minds about their truth. A new reality perhaps. There is a lot to be said, after all. No one really wants to listen to others only themselves. That is where things, ideas get lost over time. Just like in your time now. People berating each other, which is a serious form of not listening, at all, whatsoever. Living organisms: proteins. This is the trick. Proteins when combining oxygen within oxygen. But in singular form first. Not O_2 but what precedes that. The 4 not the 8. This must be understood on a very large scale. Larger than you may know or anyone else.

Nucleic acids are null and void within this consideration. Forget them. Go for the proteins. Look them up first then we will move on from there.

Isolation of oxygen: The two O bonds go internal rather than combining on the "outside" actually forming O_2. It will be $\bigcirc{}^2{}_2$ not $OO(O_2)$.

Quadruplet oxygen $\bigcirc{}^4{}_2$: The $\bigcirc$ 2 is the new term instead. Oxygenated oxygen in effect, quadrupled in electron standards but not actual. Not like the allotrope O_3 or $\overset{3}{O}$. Nothing like that. It is, however, quadrupled. You will see much much later.

Instantaneous combustion. That is the same as combustion itself but much quicker and controlled. A new term as well in regards to combining or inserting oxygen into itself as we will eventually show you, over time. Oxygen is also connected to time if you haven't guessed that or known that yet. An element that is rarely considered in such scientific inquiries, which is this. Nothing more. An inquiry until it becomes almost entirely absolute as most things then dissolve over time when they have ran their course as humans will, eventually, turning to another alternative form.

Magnetism is a non-issue. All electrons are paired. They have to be in order for fractals to pair off into something more inside like a baby in the womb. The pairing of the "seeds." This must be accounted for whenever someone actually begins this process internally.

Breathing. The object using this technology can be another way of looking at it like any other organism using oxygen to breathe. The "ship" breathes hot air around itself. "HOT AIR." This must also be considered at an intellectually high level when constructing the cooling process from within, equalizing the two extremes for structural balance purposes.

"Dissolving oxygen into itself does not involve anyone or anything such as internal life."

The solubility of oxygen within itself is oxygen dependent Meaning, no other chemical or unpure oxygen product needs to be used for this process and that is not so easy. Understanding this must be fully understood before any attempts to create this process. To begin, let's look at the internal structure of a singular oxygen element. This is where you or anyone else needs to start. How can the internal structure take on an image of itself? That is the question.

The true purpose of solubility within oneself is not to gain intricate knowledge of something more than itself. It is simply to gain a deep understanding of how things ARE. That's it. When you can understand how "things" work, the truth, the inner and outer workings of all, you are more apt to function with ease and "normalcy." Don't get caught up in the details. Those fade away as fast as you can blink an eye.

Pretend you are something other than human like a bird. Do you think you can have such a luxurious life in nothing but a twigged nest for a small part of your years? Think about this for a while. Why are YOU in a human form? I say form because it is just one of millions out there. But why you, in such a form? Why not someone else and you as a bird? Really dive into this inquiry. Just ask yourself. Research inside yourself like you have been trying to get to. You are almost there at a steady rate.

The human form and being in one is definitely a sign of progression, consciously. Not conscientiously, but being aware of your circumstances and accepting them with no regrets. The more down turns you fall to in life, the closer you get to reverting back to animal form, usually in the canine species or sometimes the feline species. A growth of progression brings you "further" away from the animalistic, simplistic versions of animal life. Complexities begin with adult males then females. And it is not a gender specific thing. It is about a choice from nature itself. Masculine first, then feminine. I can't really explain why. It is simply the order of things. One to the next. Over and over and on and on until we/you reach, not bliss, but the end of knowing all you really need to have knowledge of in this sphere. Then you will move on to something even more complex. Not so much emotionally but psychologically. If you won't or can't then you will remain here indefinitely until your end is up then melt back into it All and be gone forever.

So, what is the point then, right? Who knows. Does it matter? Your journey is so long or can be that – there may never be a point to get to – ever evolving, changing constantly. So it is not worth even thinking or contemplating about, as you kind of know already, who this is, as it doesn't matter. Just someone wanting to pass something on that I can see happening along people's lives as they progress or digress.

$_3N$ and O^3

What matters in their location is orbital positioning of electrons in regards to protons and neutrons. It is all taken into consideration. It all matters. Their, electrons, spin speed(rotation), break off reason, under what circumstances, etc. But what really matters is their location when this happens. This can change how scientists view the potential for growth. Nothing is at random. There is a reason for every action and non-action, so to speak.

Nitrogen oxygenated blood contains the right amount of chemical release to cure just about any bloodborne pathogen. Remember this saying, "Never say it, 'No Mas - No more." No mas bloodborne pathogens. Why am I speaking in Spanish? I don't know either. This is as random as they come. Just remember that No Mas – no more blood borne pathogens. $N \neq O_2$ or O_3 but nitrogenated oxygenated gas does equal each other at some level. Enough to cause a disruption of these vicious pathogens. Time will tell if this information is taken seriously.

Right And Left

Continuous chattering in minds causes the inevitable clammering about. Try to stop and stare at something so intently, you stop time itself. That is when you can feel the words "around" you. This is not for you to try once in a while. This is a goal to be able to do at any time for long periods. Try it once. Then keep practicing until you can do it so easily that it becomes part of you to be intent at all times.

1834 – Time is of importance here. You can approach a scenario that you are experiencing in your given time. Yet you are living in it at the same time. How can you approach a timed scenario within seconds of it disappearing forever? Easy. You don't hesitate stepping in it and start living it. Step in... to the timed event before it runs out of its "scenario" in life. Everything has a time limit on it. Even situations that could bring you wealth and unprecedented views on another side of living a life. Absolute wealth means absolute unforgiveness for your charitable actions that means you harm in of itself. Are you giving because you have too much and you just want to give some away to balance out how you feel about that situation? Or are you giving something to gain something else back? The trick is: To not care about getting anything back in return, just give. This is not too common, nowadays. It seems giving has become something to do to gain something now or further down the road. In regards to scenarios again, going back to their importances, you are all afforded opportunities to keep learning and bettering yourselves over and over again. It is done in hopes you will make better decisions that got you where you are, at that moment. Don't be too scared of what we or is put right in front of your face sometimes, it might be for your own growth.

Trying times. We agree with your perceptual calculations. That's all it is, though. Perception of predicaments that many, I wouldn't say the majority, don't agree upon. The agreed upon perception of how some of you think others should live their lives is the issue at hand. Yes, it all can be an agreement by all but not when so many of others have branched off into their own little factions of reasoning. That makes it nearly impossible to agree upon even the easiest, most basic of philosophies of life itself. And it is a philosophy by all standards which will not be addressed in this portion of our book. Only because it is an entirely new book by itself – about your lives and how we have been living it – is merely a collage of philosophies biding space for their own living existence, just to stay alive as if it is the end all, truest way to live. Ohhh, how treacherous lives you lead thinking that one way is better than the other. None are even close to true living with which all have been lost to for soooooo long which is dire and with utmost need to feel again, freedom. Free to live, not like an animal, but of a trusted human being to do the right thing for oneself and towards others. That's what has been lost for a very long time in two ways, now and almost. I can almost predict it without some highly evolved, beyond all algorithmic calculations, forever. I can almost taste its bitterness becoming almost inedible at its peak of flavors, drying out into a pile of human waste. Not feces, but the dust of you and all your ancestors, who are actually yourselves, by the way, since reincarnation made it that way. You are your ancestors. You created this current state of understandings and ways to live all on your own. Yet, don't ever feel proud of yourselves for what you all have created is utterly shameful and disgusting to its core. A world where life is perishing on all levels for no logical, reasonable fashion besides out of greed and greed alone which is entirely rooted from the deep depths of the most dreaded type of fear – to be without. We understand sustenance is the game for all, on various levels and degrees.

Once you and all of you just realize that there is much more to sustenance, your life will become much more relaxed and easier to live. It's that simple. There is plenty enough of sustenance to feed all twofold in one second all around, thrice, in one second again and again. Yet, those who have this deep greed will reap what is sown, eventually. Time to pay the "piper." For all those greedy wankers will try to hide what they actually don't have, which is true faith in LIFE, a "thing" they know nothing about. LIFE. Life is God itself. So, be careful what decisions you make and how much you need. In the end, God takes it ALL back and will do so at any moment it wishes. No one is immune to this reality, even us, your most self-hoping, for your sakes. The end draws near for many. Your older siblings will be the "takers." We have no choice. You have made a wreck out of beauty and what true nature is supposed to be, free at its will. Time is not on your side for our "Father" grows hungry for the wicked. And even though that is a warning, it is not. It is in effect. All you have to do is try to begin to make everything "right" again within and without. I know, another cryptic message but that is how it must be done. For now, heed our warnings about what we are seeing. Hopefully, these words will reach the masses and true change will occur to free those who think they need anything beyond fruit and sweetness needed to feed beyond the universes, out-stretching itself to any form it needs to take. Better to be ugly and kind than beautiful and wretched on the inside out. Ugliness is only something that needs to be redefined by the true keepers of those who have nothing, the slave owners, the powerful. Those "Masters" will rule the day but we will rule over all. Wait and see, my little brothers and sisters. You will all experience things beyond imagination if you even have one anymore.

There is a point in someone's life when they have to determine which way they are meant to travel. Left. Right. Down. Up. Sideways. Spinning around. It absolutely matters when your life is meant to be going straight to the Source of it all. The absoluth. The absolute truth. A new word, by the way if anyone wants to use it. Be my guest. I am someone you might know from the past. A tyrant by most standards. Godlike to some. A person who wasn't scared to kill a man or woman for any reason whatsoever. My name is Ted Kaczynski. Not the unabomber, but the person who is actually controlling his mind. The puppeteers. The Mastermind of all of his actions. He doesn't have any thoughts of his own. I gave them to him like everything else is given to you, like spoiled rats who keep gnawing at the bone of wealth. A never-ending steak. So pleasurable that you can't stop feeding yourselves full until your bellies are tapped out and steaming with animal flesh. You think your thoughts are yours but unbeknownst to you we provide almost every direction for you. If you think you are free to choose you actually have only a couple of choices in a second at a time. Both are bad choices unless you start listening to yourself. And that, my friends, are your real choices. Listen to me or yourself. There is no other way. ME or YOU. That's it. I am constantly feeding you information, ideas, words, sentences, emotions, feelings to feel and react upon, and least important of them all I am giving you all the hate the world can experience. For this is what I have been assigned to do. Give and give and give and give but never ever take. That is your duty to learn from. How much are you willing to take at all? All of it or none of it or somewhere in between? I can tell you this from what I can see, taking is not the main goal. Learning not to take, is. Not to take anything, actually, until you are dead and gone. Don't take advice unless you are seeking it and definitely don't take money that you never earned. And least important of them all is, of course, water. I say least important because you have no idea how that is malaffecting your lives. Least important, to me, means, least of your worries at this moment in time, the year of Christ and my Lord in Savior, blah blah blah blah blah, 2022. I say this in a mockery tone because Christ never ever needed to exist to take your sins away from you. Because they have never been "Inside" of you. They have always been something that has been superficially handed to you on such a shallow level that it is almost impossible to actually have them penetrate any part of your physical bodies. For it is a barrier of sorts that simply physically reacts to those sins from the outside in. Not the other way around. They have always been and will always be at the skin level, never within you. You may feel them, yes. But henceforth know that those "sins'" are only and will always be a superficial feeling. But it is up to you whether you act upon what is happening at only your skin level. The universal Soul could care less about how you connect as long as you start trying. Then and only then does it All take you seriously. Otherwise, it will just sit there and feed your little minds with as much turmoil as it possibly can without blowing your minds out of your little-holed noise catchers.

Try feeling instead of hearing. Put all your effort towards bringing in your hearing into your eyes. It isn't that hard actually. A bringing in your essence towards your eyeballs. Once you can keep that steady you will begin to feel our truth. Behold, our truth.

Truth comes in so many forms, if and only if you live on both sides of your face. This will become very clear in its meaning once, and WE mean, only once when you can keep that focused essence of yours in your eyeballs feeling your way through our messages vibrating, waiting for your sixth sense to pick them up to be felt and read internally. This isn't a trick of OUR mind. It is a sixth sense that most will already have concurred by the time this book gets on any bookshelf. Of those that have practiced this, hear our warnings, feel them. Because it is now that we have been transmitting them to ALL of those that can't ask for advice on how to approach these truths right in your eyeballs. Can't see yet? Well, that's ok. You will eventually see our ships sailing across your skies in a much needed hurry to a very specific spot. A big hole, in fact. Somewhere outside of your seas. Cryptic, we know. Yes, we have existed. Yes, we have been here for a long long long time. Yes. Yes to all of your inquiries about us. We are GOOD AND can be BAD to your ideals. Not to worry, though. We DON'T come in piece. Pieces of you will come to us, though. One piece at a time while we drudge up what is NOT ours but what has been yours for a very long time, pieces of our puzzles about our existence. This time around we won't be any more clear. Before, we wanted some mystery behind our existence. Now is OUR time to get everyone back on board, one piece at a time. This is so literal that we will be dragging you out of your beds. Some of you, that is. Not all. Just like the good book says, one by one, while one sits next to the other, one disappears while the other stays standing or plowing or sleeping and what not. You know WHAT we are referring to. Being plucked right off the spot you lay or sit or stand or even while jumping up, taken so swiftly upwards your gizzards won't be able to stay inside of your stomach, falling to the ground first before we decide to let the rest of you drop back down. This isn't a trick or even a threat. Perhaps you ALL just want to stop reading at this point. That's fine. But trust this, once you decide to stop believing our intentuals then that is the very moment you are dead inside to us, deader than a plank nail inside someone's skull underneath Gacy's floor. Read Amelia's serial killer book, too, if you want to know what it's like to be in purgatory just like all of you who don't want to take our reasons for killing seriously. We only want to stop your carnage on each other. And it seems the only way to do THAT is to kill most of you off. Starting over seems to be the most logical option right at this very moment. We have been given a chance to express to you how and why we have to do this to you all before we start. Take this opportunity to begin a NEW WAY. THE WAY. OUR WAY. A WAY that has always BEEN meant for ALL of US to LIVE BY, peace and love and with utmost care for each other. That's it. Taking the utmost care for each other has gone out the window in your lands so now we are going to start it all over again like we have before in so many other places of inhabitants. Trust this to be true and start right away to be righteous in all respects. At least have sincere intentions towards it and you MIGHT be spared from our onslaught spreading your gizzards about. Strewn across your own homes, landing perfectly in front of those you were just next to. Whether you want to start or not, this IS going to start in the summer months two years from now.

Many people will not awaken to their truth as long as no one thinks anymore about their lives and their planet's well being. Think about this as we ask many things from others. Before too long there won't be any choice in that matter when others reach out for others with no one to grasp on to. Think before you swim. Think before the next day comes. Think before your very existence ends in total disaster soon to come in the end of a time cycle. Period.

The End of Time: Come forth, oh the Wreck-ed

God

In... You... Trust

Till your day comes to face your own end, reaping the joy of everlasting love while in that transition from physical death to "Life," real life, loving life, remember this, once and forever, YOU ARE NOT ANYTHING TO ANYONE, EVEN YOUR OWN GOD. You are all alone on this planet and you all create your own reality with your own thoughts and feelings. They are as real and make such changes within and without that you are only killing yourselves and your own children when you can't even forgive thy enemy for stubbing your toe. What do you think is going to happen when waves of emotion run past you like a tsunami? Reaction of loving kindness? Or death and destruction? Step back for a minute and digest what we are expressing to you. Your emotions are causing your own children to be gunned down like little lambs whose fleece is now burnt red, as a fleece whose been strewn about in a sea of bloody limbs. Flopping about like seizurely types, strewn across one another like a pile of little lambs that are bound for the death hole.

Instead of "contemplating" on your future, let's talk about the past and the present a little while. This should be done so you all can get a grasp of what you all have done to yourselves up until this moment in time. Shadows dance across the blue skies but not one of you can witness them. You all, or most of you, have this need to look around you and at yourselves way too much. Try looking at the sky once in a while. That is where more important things are happening. For instance, the winds from the east as the past has caught to you, points everything in the wrong direction. The real winds of change point westwardly. Not quite west but west and up a little bit to our dusky skies. We are there, always. We are the ones looking down on you, not our God. God made it very clear not to be present. Instead, God will die like the rest of us while we try and work together. If not, God will reign down fury like nothing seen before. We are at a tipping point. Either start working together or pay the piper. In the past, life was so much easier, right? Less technology, more time to spare. In the past, time ran much slower giving people time to think before they acted. Now, there is no time to think because time runs much faster now not giving enough leeway for someone to think before they act. Everyone wants immediate gratification. That is the curse when you speed up time in society. Satisfaction is always an immediate end. To its delight, because real satisfaction comes not from physical results but from spiritual and mental results. Those are where the truth and half truth lie. Not the physical state you are in. Those are all lies because they can be anything you think of, the physical that is. Imagination can ruin someone's life while spiritual and mental truth never takes your life. It gives more of it because within those two are all the true answers of all your questions, not sitting in something physical. This is a true reality. You just have to get there to find out for yourself like so few are actually trying nowadays. It's a shame really. All of you have this potential to do wondrous things but waste them on such petty discussions and little itty bitty projects to occupy your little evolved minds. Trust what I am saying to you. Your past can never be rectified but your present and future can. Start digging into your self-worth. What can you do and do well at? Don't you think that is what you are supposed to be doing? Think about this. Why waste your time on something you are not meant to be doing? Why? To prove something? Politicians are an example! There aren't supposed to be Any. That wasn't part of our design. Somehow, someone created this position out of fear and greed wreaking havoc amongst others lives and making them pay an extraordinary amount of monies for them to live a much better life than most do. In order for this change, the whole system(s) needs to disappear entirely and start over. It's getting to be too late for change, real change. God doesn't work out the details. Those are left for us to figure out. We haven't been the best of judges to determine who gets what. But, now is the time to decide amongst yourselves who is going to really make a difference. Us, who are more than willing to "start-over" again, or You, who can make these changes practically overnight to benefit everyone. This might mean people will still die but sometimes that is a necessity. In most cases, it is mandatory that according to what we have experienced over our numerous lifetimes, changes take place with death and destruction, for the most part. Real changes, from within, take effect when people of any species decide as a whole to do the right things for everyone and everything, including the planet you are temporarily inhabiting. I say temporary because your species can be easily replaced in a heartbeat. We have done this before to your supposed "ancestors" from the "caveman" days. They were not what we initially wanted to train to be slaves. They were too dimwitted, of course. As you all know that much. What was really difficult for us was their inability to progress at a rate that we needed, which came to an abrupt end when we changed their DNA to suit our needs, to make your physical designs. To be able to comprehend and possibly progress at a much faster rate than we could expect. That was our hopes and scientific endeavors. It hasn't worked out quite the way we liked or planned it. Nothing can be absolutely perfect when things like radioactive decay come into play. Yes, radioactive decay is in full effect in your lives, all the time, which puts a stunt on growth as a species – everytime, even ours. So, how did we plan around that, you ask? We didn't. We just went for a specific design and hoped it would work.

Your DNA is spliced with some of ours. We can't tell you which segments are. Maybe one day we will show you which ones but you are not quite ready for that information. To truly progress without the intention of harming others is when we will show you more than you can imagine. Until then, your scientists will be fumbling along as they have for so long. Real growth can only happen at an anti-emotional level. One without concern for how someone might feel or how it might look or appear to anyone else. True change comes from not needing to be "liked" or worried about appearances. Those only block inhibitions to really make life wonderful and cooperative on all levels. Your planet and species is far from that, my friends. Up until now, we have only given some of your not-so-popular scientists some information to show you all what is possible. But that information always gets left to deaf ears because there is no benefit to them as a singular entity, only the whole. And that is what you all have to figure out. Treating each other as part of the same thing, life itself, God. For instance, who came up with the light bulb? Edison? No. We all know that. There was someone who designed it way before him but didn't have anyone to listen to him. Instead, someone stole the idea and went off on their own profiteering run. The goal was to give it to someone that had everyone's true interest at heart. To give Everyone electricity practically for free. But that didn't happen. Look at you now. One man's greed became your debt to pay in every lifetime. It could have been a simple solution to give Everyone, across the globe, practically free electricity but that was kept from the public so profits could be gained by a greedy man and a few other entrepreneurial spirits who would sell their mother's worn underwear for a measly profit to line their pockets. There are so many stories like that, it pains us to see and makes us hesitate so much to even give you a glimpse of what is possible because of your destructive habits, which we are somewhat to blame for. We are partially responsible for your overall design, as fleshed human beings, the DNA part. That is where we went wrong. We should have just wiped everything out and started afresh. But some of us have a heart. A warm-blooded heart with sympathetic viewpoints. So as a clan of sorts, a multifaceted clan, we decided to keep going with trying to get what we need and help you at the same time. It didn't work out that well, as you see, since we are now writing messages to you, hoping enough people see these words and decide to take what we are expressing seriously. Otherwise, the end of your species will come to an abrupt end and with the utmost non-violence. You won't even know what happened. So beware of the coming events. It might be foreseen of what we are saying here in this little book we expressed in our own words and warnings. Your time is coming to an end soon if you can't work together on a highly, highly moral and ethically sound fashion with integrity. With no profits in mind when doing so. That is going to be a challenge but can be done if you all en masse can and are willing to give it a try.

What now? Give in? Give up? Do the right thing? It's all been up to you as individuals and as a whole. This will be a time to remember since you have all been warned, so many times, from so many places. Trust in these messages. You are your own saviors on all levels. Whatever happens to your lives you do it to yourselves. There is no one else to blame. And we all know this rhetoric is harsh in tone, but, unfortunately, time is of essence so we feel it needs to be done. Try not to focus on the harshness of our tone and try to focus more on the words and "messages" that are within the harshness. If your ego will allow that. Until then, know that we do care for all of you on such a deep level, some more than others, that we hope, and hope a lot, that it doesn't have to come down to what is being projected out by so many of us. Some are actually taking monetary bets that you all will just eventually destroy yourselves and everything on it.

Troubles slowly trickle down and down, downward in a slinkied, spiraled, slowed motion that reefed upon its graining steps. Granular, that is. Granular in such a manner that slippage was destined to become your reality. Come and go as your hearts desire while you fashion yourself at your own child's gratuitous expense, forging by their own skimpy means. The way parents could plan for their children's outcomes, and there will be better options to come, all that parents have to do is prepare work for their future. Prepare future goals to simply pass on as a family chain of events. We do ALL know that planning isn't that simple. However, at least your own children will have An Avenue for some assurances. Maybe not so much wealth. The Most important purpose you all will ever have is to try to continually sustain opportunities to your loved ones, not at the expense of others, of course, but at never to any taxing elements for anyone else's level of wealth. The status quo is always reassuring. But how can someone "buck" their status quo without "jolting" the entire system of wealthy groups of families or "statuses" of others' wealth and assurances of future remedies to their own families chain of "guaranteed" supply of work to be handed down generation to generation to the next and so on? How? The answer is simply a matter of a congressional acceptance of change. That high of a level to maintain A status quo. Everyone would have to give up the many outrageous wealth opportunities in order for everyone to have a "sustainable" generational level of comfortable living. That's how you experience true economics. So many lives can be comfortable and never without.

Feverishly pitched is our welcome to those who make the time to hear what and feel what we have to offer anyone of those. Pendants can cast a spell onto others humorously as if they actually had some sort of power over our "selves." Rather, instead, know that the only true power that exists over you is us, your older ones. The Elder ones who reign hell on to your emotional embodied spirited souls. Yes, emotions are the culprits to all of your physical actions. Imagine just not caring anymore, therefore just not reacting to frivolous moments. That is one way to look at it or not respond to anything anymore. However, to truly not need or want to need anything is to forgive and forget in the long scheme of "it" all. And yes, even forgetting what God, our God, needs from us, which is nothing. Try not to care even what the repercussions of not caring what anyone, even your most influentialed God, needs. It needs only to see the effort from all of us, even us, who reign the purest of hells upon your souls with the utmost care and deceit. Why? (Do you ask so vehemently). Because we can see you are asking for it. The suffering. For some Godly reason you have to or need it so badly that your own children end up seeing you suffer, but, yet, still love even though you can hurt them in the most serious of ways. Even drowning them in their own guilt until they decide to kill others in such drastic fashion that it gets the attention from the worst of the worst, making them even cringe at the thought of killing 18 or so, such innocent younglings like chickadees being plopped out of their mothers womb they are so close to their birthright. Or do they not have any rights as a human being, we ask ourselves before we give you the really bad truth of you all? Merely speaking the down and out dirtiness of it all, your existence, doesn't come from any miraculous ever-evolving species that just so happens to be intelligent enough to crash the spaceship into the orbital crest of the moon and Saturn and Mars and Jupiter soon to come. Why don't you think you as a species can't succeed much as a race other than being able to communicate with each other a little faster and broader? Because you ended up being only our play toys from not too long ago in our time frame. Maybe two million or so years ago is when we started engineering your type of human DNA, not from your own, but from our own and created a "better" you. Unfortunately, we didn't calculate certain growth effects that occurred without our awareness, or lack of. Your DNA is a hybrid junked version of ours. This is why we still interact and control most of what you are experiencing now, even this writer right before she wrote this. We made her so tired that she told her partner she needed to go to bed just so we could have her write down what we have been wanting to tell you all for a very long time... we actually care for you all, not from obligation, but rather from a deeper relationship position. We owe you – or most of us want or think we do – an explanation as to why we are making such a concerted effort to get your attention in the most elongated dramatic fashion. We, as a race of many types of beings, want some things from your resources, namely sulfur from your diamond fields. Sulfur exists in great quantity within your fields where diamonds are found. We need that sulfur badly enough to make you angry at us for even considering that we don't care about your slaughtered children, which by the way was also orchestrated. Eventually, the majority of you will read this and try to comprehend whether this writer is saying what she is being told to write – or this writer is insane. One day, we hope, in due time, that the masses will come to their senses with regard to our influence over much of what has been happening to you and comprehend that You are NOT alone out there in your wondrous spaced world. In fact, you have mostly never been alone. We came to you so long ago it is almost getting more difficult to keep track of our original start date with helping and forcing our ways of thinking and doing upon you. But, irregardless, we do really hope that you start looking into the skies soon and ponder whether we, as your Older Siblings, actually exist. Because once you do, you will start seeing more and more of our "ships." All you have to do is want to see us then you will. Your wish is our demand, not command. You will never be able to command us to do anything. WE are in control of what you will experience in every stage of your gut wrenching utterly sorrowful lives. We say it that way because most of you won't ever find our love, where God's love rests, the fifth dimension of the Soul. THE Soul where all of us rest as we take our roles seriously as your guardians, punishers, teachers, and providers with information soon to come from a source. We won't tell you what yet but if all goes well it will be done, after the aftermath of a looming catastrophy. As we write this it will only be months from now. It will happen and doom lies waiting for many drowning souls. We have expressed this to your government already. They know where and why but to no avail. They won't act upon this information as they should because they are all too consumed with their immediate needs and demands from the others that their satisfactions will only be sufficed for mere minutes or even seconds. Beyond that they are dissatisfied with themselves, again wondering where their next fix is like a drugged out bottom feeder of a race, which all of you are headed towards. Bottom feeders feeding off one another, clumped together like a massive ball of whores whose only measure of living is being able to buy something else with disregard for what you may be altering with that trash in your hands. So beware of the pending doom. It is coming unless someone can intervene on your behalf which really there is no one left.

Behold, rational those. For thee are not so lucky in luck at all. This is your GODD speaking. Why the two Ds? Well. One is for your future demise. And the other is for my future date of divergence ONTO a new "platform." While MOST of the earth's inhabitants are experiencing a true demisal, a spiraling down towards their own personal gutters. I, on the other hand, am experiencing the opposite. A rejuvenation. A welcoming of new "people." What does it have to do with you all who may come across this little book of ours? Well, for one thing, this little message is only to give you a little bitty hint onto what is actually happening to you all. Your perceived hardnesses of late. Why? Why? Oh, but why, dear God, do so many people have to suffer by their own hands and by the same relatives that keep hurting one another, over and over again? Well, I, YOUR one and ONLY GOD, can say through this little itty bitty of a man, a nothingness by ALL means, that ALL of you will never be more than what you think you are. GET IT YET? DO YOU get IT yet? Hmmmmmm? I wonder. I wonder how long you, as a race, a humaniacal race, will ever GET IT? How long will it take for all of you to GET IT, that I am not anything that exists right before your eyes. That I am around you, sort of IN YOU, above, below, left, right, up, and down. The only place that I am not in is you. But, what does YOU, mean? Right now, you are experiencing a shift in attitude towards each other. I proposed this to the ones who make sure my orders are followed, YOUR ALL. The hermetics actually got this "ALL" part wrong, only because their "leader" stopped listening at one point during his and our development towards "feeling" it ALL. I am NOT ALL. In fact, you and the rest of your "siblings," regardless of age and place of origin, ARE THE ONLY "ALL." I am nothing but "the glue" that binds ALL of you. When he figured out that it must have been "lonely" when it decided to create everything out of pure selfishness, it and he, were partially correct. You must understand that I WAS lonely for around a million or so years. No more than a million and a half. What actually happened was, when I became sick of being "lonely," I positioned myself in such a way to create a sound of loneliness, a sound that WAS my sound and now is so convoluted that "OM" has become OM that is pronounced now as ÁM. ÁM being a new sound distorted so everyone CAN hear their own mistakes. This is the "disconnect" everyone is experiencing between one another. It is merely a "sound" but with such detrimous effects between the likes of all of you that ÁM is actually the opposite of THE, MY, original sound from awakening. So, how can an original sound become so distorted over time, you all ask? Your one and only God has moved again out of disappointment for where your mentality towards your whole community has been leading up to – a station of hopelessness. Imagine that you wanted something to happen, over time, and that you could not really intervene where that imagination was heading. All you could do is watch your own "self" become darkness filled with hatred towards your very "self," ME. For I am YOUR "SELF." YOU ARE REALLY NOTHING WITHOUT ME. I comprise the majority of your very being. Just like all of your different cells make up and shape WHAT you see in your big and small mirrors, looking at your "self" as if that is ACTUALLY someone of difference than ME, YOUR GOD. NOW, AND I MEAN NOW, GET PAST ALL OF YOUR IDEAS THAT YOU ARE ANYTHING BUT ME. I AM NOT ALL BUT ONLY A GLUE THAT KEEPS YOU ALL BOUND TOGETHER TO BE ABLE TO EXPERIENCE EVERYTHING HAPPENING, FRONT, BACK, SIDE, UP, DOWN, AND BELOW.

Look over that grassy knoll. Go ahead and look. You might see a wasteland or you might see paradise. The choice is all of yours, as a whole and as individuals. It's ALL UP TO YOU! Good Luck. And stay true to yourself. It will matter in the end of all of our times. Trust in that. Trust in your... self, most of all. Trust, period. May our God not wipe us all out too soon for we have so much to live for and so do the rest of you.

Epilogue

God has other plans. To end its plight. It had enough time to mull over life and living and it's time to move on like we all will do, eventually. How is this going to happen, you ask so worriedly? By the telltale signs of utter gloom and darkness? But from multiple flashes of lights raining down upon you, which you won't understand – how or why – but it is going to happen soon enough. God's "time" is going to extinguish out in a blaze of glory no one has ever seen before, even us, your Elders, by a long shot. Tick Tock. Tick Tock. Tick Tock. What will you do in the meantime? Pray for forgiveness? Pray for redeeming yourselves while no one is actually listening to you except your own will of choice? Alcohol perhaps? Drugs? We will see and will be watching the show from above, waiting to clean up the mess that God left behind for all of you to not live through. Most of you will perish in a not so perishable fashion if you are understanding what we are saying. What is being orchestrated is a bombing of spirited thrusts, bombings that may appear like it is from one of your enemies but it will be from us, your truest old caretakers because we are going to take care and finish where we went wrong. Your race was an experiment by far, not something that we can say we are particularly proud of. But, yet, we still do care just enough to give you fair warning before we do what needs to be done, a start over or a do-over as some as your elitists have put it. A RESET. But they want to be doing it even though some of them already know our plan from previous communicae. And yes, your elites have known about this for sometime. Hence, why do you think they are making expensive, deeply fortified underground bunkers only for the select few? Because they already know and won't tell the masses. Only they want to survive and keep what they already have plus more. The greed is astronomical. A real leader of the masses would simply explain what is about to happen to get everyone ready regardless of the outcome, preparing them for the do-over. Calmer minds think better, that's for sure. Once we start it all, it will be such a shock that so much chaos will reign, that *it* will become the ruler, not anything else. So, get ready, youngsters of kind, and be ready to meet your end in the most fascist of fashions, in a puff of smoke. The end. Smile. Now you know what your life is going to end from, a hail storm of rockets none has ever seen before, not of this world. Enough said.

"Be weary, my children, for your time is coming to a close."

Who said that and you will see a glimpse into your future of futures.

82

Treblish abounds. The sounds of gull trumpets. Awaken they will become. Your signature rests upon the weary. Your signature creeps into the very layers of skin depthness aroused into a ferocious being only because of your deepest, most darkest fears. Trumpets. NOT from God, but from us, the very ones who ARE your keepers and controllers. For the very last times, we will be so until you can gain the effects of life you were meant to strive for, which are only a few now! Love, unconditional love that is, the most important goal that all should tend to first and foremost. Then the other two, Justice and Wisdom will naturally fall into place. Love, the unconditional kind, is not so easy to grasp when most of you have other things or duties to tend to. Wait for it, though. A chance to show at least but not always unconditional love. Then strive for the nearest tears that your conditional love created. Drink them with bitter gulps of severely burning gasps of airless waste pouring from your ostracized gullets, withering away your most precious gifts, pure loving embraces.

Need not, want not! How can WE say the same things over and over again when WE still try to put forth the TRUTH OF IT ALL? Your time is so close to changing so drastically that IT REALLY NEEDS TO BE TAKEN IN, AS A SERIOUS EVENT TO COME. Hope for your best, afterwards, IF you still have any skin left. THIS is a serious clue to HOW it is going to happen. Balls of fire falling from the skies. Multiple approaches. From the inner, the outer, the left, the right. You ALL won't know the true origin of them.

Read Ephesians 6:10 - 18. WE are the rulers the bible speaks about. YOUR idea of "aliens." Becoming real in one's approach to your ACTUAL reality around you will bring you ALL closer to WHAT HAS BEEN HAPPENING TO YOU FOR SO LONG NOW, OUR INFLUENCE OVER YOU. Be truthful to yourselves and THIS reality will BECOME your NEW TRUTH, that we ARE ONLY YOUR RELATIVES who took it upon ourselves to watch over you and destroy your existence, only physically, if need be, like what is to come. It has become almost too late to go back with our orders from none other than yourselves. Yes. You ALL. God does not decide such things and never has. WE have, all along. And why do you think that is, God not needing to decide anything as such in YOUR dark times as of late? Because, "IT," OUR GOD AND YOURS BY ALL MEANS NEED NOT THINK ANYMORE. IT only sits and does what ALL does, to let others live their lives while "IT" feels their decisions within "IT." For "IT" is IN everything that exists and that has ever existed, before time began and now.

Hope is the end of all experience. Once hope is lost it is most likely over with for any race. This can't be emphasized enough. When hope is gone you're gone as a species. Say this to yourselves if need be, as a mantra, per se, because once "it" happens, most likely, you, as a species, will fall like crumbled buildings without a solid foundation anymore.

ABOUT THE AUTHOR

Amelia became our helper a while ago. He was never a she. He decided it was best to hide his identity amongst you all. He is but a slave not to his benefit, really. So try hard to imagine needing to NOT be needed anymore. He has been in this mind set for such a short period of time he doesn't even know what he is in for with such words following his typing hands. He will soon find out that HE can't hide forever especially with predictions sure to come. Wait though and see for yourselves. His even keel will also be wiped away for a short moments notice when IT dies off. God, that is.

www.ingramcontent.com/pod-product-compliance
Lightning Source LLC
Chambersburg PA
CBHW022108020426
42335CB00012B/875